Co-operative and Energy Efficient Body Area and Wireless Sensor Networks for Healthcare Applications

Co-operative and Energy Efficient Body Area and Wireless Sensor Networks for Healthcare Applications

Akram Alomainy
Queen Mary, University of London, UK

Raffaele Di Bari
EADS Astrium, UK

Qammer H. Abbasi
University of Engineering and Technology, Pakistan

Yifan Chen
South University of Science and Technology of China, China

Editor
Lorenzo Mucchi

Academic Press Library in Biomedical Applications of Mobile and Wireless Communications

AMSTERDAM • BOSTON • HEIDELBERG • LONDON
NEW YORK • OXFORD • PARIS • SAN DIEGO
SAN FRANCISCO • SINGAPORE • SYDNEY • TOKYO

ELSEVIER

Academic Press is an imprint of Elsevier

Academic Press is an imprint of Elsevier
32 Jamestown Road, London NW1 7BY, UK
The Boulevard, Langford Lane, Kidlington, Oxford, OX5 1GB, UK
Radarweg 29, PO Box 211, 1000 AE Amsterdam, The Netherlands
225 Wyman Street, Waltham, MA 02451, USA
525 B Street, Suite 1900, San Diego, CA 92101-4495, USA

First published 2014

British Library Cataloguing-in-Publication Data
A catalogue record for this book is available from the British Library

Library of Congress Cataloging-in-Publication Data
A catalog record for this book is available from the Library of Congress

ISBN: 978-0-12-800736-5

For information on all Academic Press publications
visit our website at **store.elsevier.com**

This book has been manufactured using Print On Demand technology. Each copy is produced to
order and is limited to black ink. The online version of this book will show color figures where
appropriate.

Working together
to grow libraries in
developing countries

www.elsevier.com • www.bookaid.org

CONTENTS

INTRODUCTION

According to Media Lab, MIT, USA, *"By 2015, wearables will have virtually eliminated desktop, laptop, and handheld solutions altogether..."* [1].

Communication technologies are heading toward a future with user-specified information accessible at the fingertips whenever and wherever required. In order to ensure smooth transition of information from surrounding networks and from shared devices, computing and communication equipment need to be incorporated into our daily clothing. Body area networks (BANs) have been receiving an increasing interest within the wireless personal communication community. Mainstream wireless systems' designers are targeting BAN mainly for medical applications, such as health monitoring sensors which enable monitoring patients wirelessly, targeting their locations and freeing them from being constrained to a small restricted area. In addition, personalized entertainment technologies and also the pleasure of having powerful computational tools available instantly have encouraged researchers to widely study body-centric networks from various aspects. With the advances in small and low-cost radio transceivers and RF front-ends development, the possibility of applying ubiquitous and noninvasive sensors integrated into user's daily clothing, and living activities seem more feasible. The ability to share data increases the usefulness of personal information devices, providing features not possible with independent isolated devices. Current wireless sensor solutions are limited in that they do not provide the means to overcome obstacles and shadowing of propagating radio waves. Thus, for reliable communications, an increase in power consumption is required, reducing battery life. In this book, we address these limitations by designing efficient and compact antenna systems. The system will be cooperative and also aware of the surrounding environment and neighboring units and thus provide efficient and low-power wireless connectivity for personal area network and BAN applications.

CHAPTER *1*

Introduction to Body Area and Wireless Sensor Networks

Body-centric wireless networks (BCWNs) refer to networking over the body and body to body with the use of wearable and implantable wireless sensor nodes. This subject combines wireless body area networks (WBANs), wireless sensor networks, and wireless personal area networks (WPANs) [2]. BCWN has got numerous applications in everyday life including healthcare, entertainment, space exploration, military, and so forth [3]. The topic of BCWN can be divided into three domains based on wireless sensor nodes placement, that is, communication between the nodes that are on the body surface; communication from the body surface to nearby base station; at least one node may be implanted within the body. These three domains have been called on-body, off-body, and in-body, respectively [2]. Figure 1.1 shows an example of on- and off-body systems only; for in-body communications, one of the nodes should be implanted within the body.

The major drawback with current on-body systems is the wired or limited wireless communication that is not suitable for some users and the restrictions on the data rate (like video streaming and heavy data communication, where we need to transfer a large amount of data). Many other connection methods like communication by currents on the body and use of smart textile are proposed in the literature [3], but communication by current method suffers from low capacity, whereas smart textile method needs special garments and is less reliable. BCWN seems to be the most suitable communication method because of the less power requirements, reconfigurability, and unobtrusiveness [4]. However, in order to make these networks optimal and less vulnerable, many challenges, including scalability (in terms of power consumption, number of devices, and data rates), interference mitigation, quality of service, and ultra-low power protocols and algorithms, need to be considered.

Co-operative and Energy Efficient Body Area and Wireless Sensor Networks for Healthcare Applications.
DOI: http://dx.doi.org/10.1016/B978-0-12-800736-5.00001-5

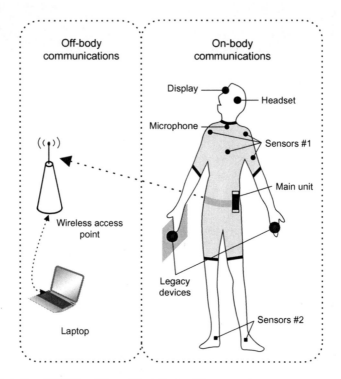

Figure 1.1 Illustration of BCWNs and its possible application [6].

Short-range devices and networks operate mainly stand alone in indoor home and office environments or large enclosed public areas, while their integration into the wireless wide area infrastructure is still nearly nonexistent. The wireless body-centric network belongs to this category of system naturally due to the small distance and coverage range required between the various wireless components. When designing future short-range wireless systems, the increasing pervasive nature of communications and computing is required to be accounted for. Developers and researchers alike assume that the new wireless systems will be the result of a comprehensive integration of existing and future wireless systems. Body area networks (BANs) are a natural progression from the WPAN concept, and they are wireless networks with nodes normally situated close to the human body or in everyday clothing [1].

The advances in communication and electronic technologies have enabled the development of compact and intelligent devices that can be placed on the human body or implanted in the body, therefore facilitating the introduction of BANs. BAN stands for wireless

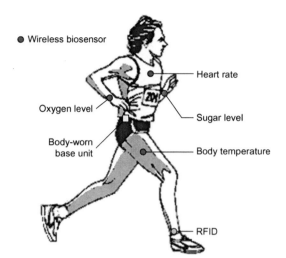

Figure 1.2 Wireless BAN and sensors network applied for healthcare monitoring applications [3].

communication between various components attached to the body, such as data spectacles, earphones, microphones or sensors, by its wireless connections between the individual components. BCWN is aiming to provide systems with constant availability, reconfigurability, unobtrusiveness, and true extension of a human's mind. Figure 1.2 presents an illustration of BAN applied in medical care services. BAN can be applied to many fields, and some of its applications include:

- Medical applications: BAN can be used in transmitting body parameters such as blood pressure, pulse rate, body temperature to main processing unit.
- Assistance to emergency services such as police, paramedics, and firefighters.
- Military applications including soldier location tracking, image and video transmission, and instant decentralized communications.
- Augmented reality to support production and maintenance.
- Access/identification systems by identification of individual peripheral devices.
- Navigation support in the car or while walking.
- Pulse rate monitoring in sports.

The ultimate WBAN should allow users to enjoy those applications and more with transmission without interferences, low transmission power and low complexity due to network organization by base station, extremely small hardware, broadband local area communication and ad hoc networking, simple channel and source coding, and scalable data rates. BANs have distinctive features and requirements that make them different in comparison to other wireless networks. These include the additional restriction on electromagnetic pollution due to proximity to the human body which requires extremely low transmission power. The devices deployed within BAN have limited sources of energy due to their compact and small size. Some devices are implanted in the body which means that regular battery recharging is not a feasible option. Due to the large number of nodes for specific applications, which are placed on the human body within a relatively small area compared to WPAN, the interference is quite strong. In addition, the human body tissue is a lossy medium; hence the wave propagating within the WBAN faces large attenuation before reaching the specified receiver.

As mentioned earlier, the main characteristics of BCWN operating device should be low power requirements, less complexity, low cost, robustness to jamming, low probability of detection, scalable data rates from low (1−10 kbps) to very high (100−400 Mbps), and very small and compact in size [5−7].

CHAPTER 2

Frequency Band Allocation for Body Area Networks

Wireless communications systems can operate in the unlicensed portions of the spectrum. However, the allocation of unlicensed frequencies is not the same in every country. Important frequency bands for BAN are reported in Table 2.1 and they are:

- *Medical Implanted Communication System (MICS)*: In 1998, the International Telecommunication Radio sector (ITU-R) allocated the bandwidth 402–405 MHz for medical implants [5]. MICS devices can use up to 300 KHz of bandwidth at a time to accommodate future higher data rate communications.
- *Industrial, Scientific, and Medical (ISM)*: ISM bands were originally preserved internationally for noncommercial use of radio frequency. However, nowadays it is used for many commercial standards because government approval is not required. This bandwidth is allocated by the ITU-R [6], and every country uses this band differently due to different regional regulations as shown in Table 2.1.
- *Wireless Medical Telemetry Services (WMTS)*: Due to electromagnetic interference from licensed radio users such as emergency medical technicians or police, the Federal Communication Commission has dedicated a portion of radio spectrum, 608–614, 1395–1400, and 1427–1432 MHz, for wireless telemetry devices in the United States [7] for remote monitoring of patient's health; however, such frequency bands are not available in Europe. WMTS is approved for any biomedical emission appropriate for communications, except voice and video.
- *Ultra Wideband (UWB)*: It is a communication system, whose spectral occupation is greater than 20% or higher than 500 MHz. Initially it was available only in the United States and Singapore but on August 13, 2007, Ofcom finally approved the use of UWB wireless technology without a license for use in the United Kingdom. More details about UWB are mentioned later in this chapter.

Co-operative and Energy Efficient Body Area and Wireless Sensor Networks for Healthcare Applications.
DOI: http://dx.doi.org/10.1016/B978-0-12-800736-5.00007-6

Table 2.1 Unlicensed Frequencies Available for Personal Networks [7]

Name	Band (MHz)	Max Tx Power (dBm EIRP)	Regions
MICS	402.0–405.0	−16	Worldwide
ISM	433.1–434.8	+7.85	Europe
ISM	868.0–868.6	+11.85	Europe
ISM	902.8–928.0	+36 w/spreading	Not in Europe
ISM	2400.0–2483.5	+36 w/spreading	Worldwide
ISM	5725.0–5875.0	+36 w/spreading	Worldwide
WMTS	608.0–614.0	+10.8	the United States only
WMTS	1395.0–1400.0	+22.2	the United States only
WMTS	1427.0–1432.0	+22.2	the United States only
UWB	3100.0–10600.0	see Figures 1.2 and 1.3	the United States, EU, etc.

CHAPTER 3

Antenna Design Requirements for Wireless BAN and WSNs

An essential part of the communication system enabling the realization of this concept is the RF front-end antenna. The complexity of the antenna system design depends on the radio transceiver requirements and also on the propagation characteristics of the surrounding environments. For the conventional long to short wave radio communication, conventional antennas have proven to be more than sufficient to provide desired performance minimizing the restraints on cost and time spent on producing such antennas. On the other hand, for today's and tomorrow's communication devices, the antenna is required to perform more than one task or in other words the antenna needs to operate at different frequencies to account for the increasing introduction of new technologies and services available to the user. Therefore, careful considerations need to be taken when designing the antenna and deploying the communication devices. This applies to antennas used in body-worn devices that need to be somehow hidden and small in size and weight. This chapter briefly introduces wireless personal networks and the progression to body area networks (BANs) highlighting the properties and applications of such networks. The main characteristics of body-worn antennas, their design requirements, and theoretical considerations are discussed with respect to recent advances within the wearable antenna research and development area. Comparing and analyzing the effects of various antenna types on the radio wave propagation within the specified environment demonstrate the major role of the antenna in shaping the propagation channel models, specifically in body-centric networks. To give clearer picture of the practical considerations needed when designing antennas for body-worn devices deployed for commercial applications, a case study is presented with detailed analysis of the design and performance enhancement procedures followed to obtain optimum antenna system for the proposed medical sensor [5].

Antennas play a vital role in defining the optimal design of the radio system, since they are used to transmit/receive the signal through

Co-operative and Energy Efficient Body Area and Wireless Sensor Networks for Healthcare Applications.
DOI: http://dx.doi.org/10.1016/B978-0-12-800736-5.00006-4

free space as electromagnetic waves from/to the specified destination. The characteristics and behavior of the antenna need to adhere to certain specifications set by the wireless standard or system technology requirements. This means that the transmitting and receiving frequency bands of the various units need to be justified accordingly. Another important parameter is the gain of the antenna that directly affects the power transmitted and since there are restrictions on the level of power that the human body can be exposed to, the design of the antenna and the other RF components require careful consideration of this issue. In designing antennas for wearable and handheld applications, the electromagnetic interaction among the antenna, devices, and the human body is an important factor to be considered. Various application dependent requirements necessitate thorough evaluation of different antenna configurations and also the effects of multipath fading, shadowing, human body absorption, etc.

3.1 THEORETICAL CONSIDERATIONS

Antenna designs and parameters are presented widely and in great detail in the open literature making the science of antenna design and analysis a mature science. Conventional antenna parameters include impedance bandwidth, radiation pattern, directivity, efficiency and gain which are usually applied to fully characterize an antenna and determine whether an antenna is suitable for specific applications. These parameters are usually presented considering the classical situation of the antenna placed in free space; however, when the antenna is placed in or close to a lossy medium, the performance changes to that of antenna placed in a vacuum which is the ideal environment, and the parameters defining the antenna need to be revisited and redefined.

In a medium with complex permittivity and nonzero conductivity, the effective permittivity ε_{eff} and conductivity σ_{eff} are usually expressed as

$$\varepsilon_{\text{eff}} = \varepsilon' - \frac{\sigma''}{\omega} \tag{3.1}$$

$$\sigma_{\text{eff}} = \sigma' - \omega\varepsilon'' \tag{3.2}$$

where the permittivity and conductivity are defined in their real and imaginary parts

$$\varepsilon = \varepsilon' - j\varepsilon'' \tag{3.3}$$

$$\sigma = \sigma' - j\sigma'' \tag{3.4}$$

The permittivity of a medium is usually scaled to that of the vacuum for simplicity

$$\varepsilon_r = \frac{\varepsilon_{\text{eff}}}{\varepsilon_0} \tag{3.5}$$

and ε_0 is given as 8.854×10^{-12} F/m.

The equations indicate the differences between free space and lossy material, hence the imaginary part of the permittivity includes the conductivity of the material which defines the loss that is usually expressed as dissipation or loss tangent

$$\tan \delta = \frac{\sigma_{\text{eff}}}{\omega \varepsilon_{\text{eff}}} \tag{3.6}$$

The biological system of the human body is an irregularly shaped dielectric medium with frequency dependent permittivity and conductivity. The distribution of the internal electromagnetic field and the scattered energy depends largely on the body's physiological parameters, geometry, as well as the frequency and the polarization of the incident wave.

Figure 3.1 shows the measured permittivity and conductivity for a number of human tissues in the band 1–11 GHz. The results are obtained from a compilation study presented in Refs. [8, 9], which covers a wide range of different body tissues. Therefore, one major difference that can be identified directly when placing antenna on a lossy medium, in this case the human body, is the deviation in wavelength value from the free space one. The effective wavelength λ_{eff} at the specified frequency will become shorter since the wave travels slower in a lossy medium (λ_0 is the wavelength in free space)

$$\lambda_{\text{eff}} = \frac{\lambda_0}{Re\left[\sqrt{\varepsilon_r - j\frac{\sigma_e}{\omega\varepsilon_0}}\right]} \tag{3.7}$$

However, the effective permittivity as seen by the antenna depends on the distance between the antenna and the body and also on the location since the human electric properties are different for various tissue types and thicknesses. The general rule of thumb is that the further the

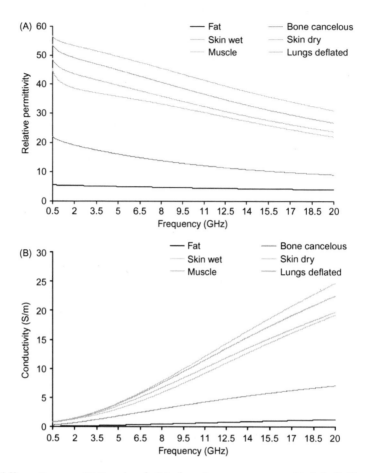

Figure 3.1 Human tissue permittivity and conductivity for various organs as measured in Refs. [8, 9].

antenna from the body, the closer its performance to free space one. Moreover, this also depends on the antenna type, its structure, and the matching circuit. Wire antennas operating in stand-alone modes and planar antennas directly printed on substrate will experience changes in wavelength and hence deviation in resonance frequency, depending on the distance from the body. On the other hand, antennas with ground planes or reflectors incorporated in their design will experience less effect when placed on the body. This is independent of distances of antennas from the body.

An important factor in characterizing antennas is the radiation pattern, hence gain and efficiency of the antenna. The antenna patterns and efficiency definitions are not obvious and cannot be directly derived from conventional pattern descriptors when the antenna is placed in or on a lossy medium. This is due to losses in the medium that cause waves in the far field to attenuate quicker and finally to zero. Antenna efficiency is proportional to antenna gain [5],

$$G(\theta, \phi) = \eta D(\theta, \phi) \tag{3.8}$$

where η is the efficiency factor and $D(\theta,\phi)$ is the antenna directivity which is obtained from the antenna normalized power pattern P_n that is related to the far field amplitude F,

$$D(\theta, \phi) = \frac{P_n(\theta, \phi)}{P_n(\theta, \phi)_{\text{average}}} = \frac{\left|\vec{F}(\theta, \phi)\right|^2}{\left|\vec{F}(\theta, \phi)\right|^2_{\text{average}}} \tag{3.9}$$

Calculating the wearable antenna efficiency is different from that in free space, due to changes in antenna far field patterns and also in the electric field distribution at varying distances from the body. However, the radiation efficiency of an antenna in either lossless or lossy medium can be generalized as,

$$\text{Efficiency}_{\text{radiation}} = \frac{\text{Radiated power}}{\text{Delivered power}} \tag{3.10}$$

A figure of merit, which is in direct relation to antenna patterns and with great interest in wearable antenna designs, is the front–back ratio. The ratio defines the difference in power radiated in two opposite directions wherever the antenna is placed. The ratio varies depending on antenna location on the body and also on the antenna structure. For example, the presence of the ground plane in a patch antenna reflects the electric field traveling backward; hence the front–back ratio is not significantly different when placed in free space and on the body, which is not the case for conventional dipoles or monopoles with radiator parallel to the body.

3.2 WEARABLE ANTENNA DESIGNS FOR WIRELESS BAN/PAN

This section presents numerical analysis of antenna performance when placed on the body and the effect of distance variation and antenna on-body positions on the main characteristics. The main objective of the study, detailed here, is to develop an understanding and characterization of body-worn antennas and the body presence effect on general antenna parameters, including impedance matching, radiation patterns, gain and efficiency. The unique on-body propagation channel between various antenna pairs is also an essential part of the general understanding required for wearable devices. The investigation includes conventional monopole, dipole, and other modified designs (all planar and printed). Transmission characterization and coupling between various antenna pairs at different on-body links are demonstrated including effects of reflectors presence on the path link and field distribution around the human body. In addition, a compact sensor design will be presented which is suitable for healthcare and activity monitoring applications [10, 11].

Different planar antenna types are analyzed and studied to investigate the effect of human body presence on their characteristics and to establish the relationship between antenna structure and interaction with human lossy tissues. Six designs derived from conventional dipole, monopole, and loop antennas are studied [5–7, 10] (Figure 3.2). All antennas are printed on an FR4 board with dielectric constant of 4.6, conductivity of 0.002 S/m, and thickness of 1.6 mm (except for the wiggle antenna, thickness of 0.7 mm is applied).

Figure 3.2 presents the schematic designs of the antennas applied with the return loss (hence impedance matching performance) of the antennas demonstrated in Figure 3.3. Figure 3.4 demonstrates the simulated antennas' radiation patterns. Illustrated patterns are the antennas' radiation performances in the azimuth (xz) plane at 2.44 GHz. In general, all antennas demonstrated excellent performance within the industrial, scientific, and medical (ISM) band of interest (2.4–2.48 GHz) with omnidirectional patterns and excellent gain and efficiency values.

The effect of human body lossy tissue presence on the antenna performance and the dependence of the antenna characteristics on distance and location from and on the body are analyzed and numerically investigated. Antenna return loss, radiation patterns, and gain and efficiency at the frequencies 2.4, 2.44, and 2.48 GHz, within the ISM band, are presented with respect to distance from human body (1, 4, and 8 mm) and also on antenna location on the body which includes left chest, right chest, left ear, left waist, and left ankle (Figure 3.5). The model applied is the commonly available detailed multilayer human model, namely the visible male model developed by the US air force (http://www. brooks.af.mil/AFRL/HED/hedr/). The cell size applied in the modeling is $\lambda/8$ (at 2.4 GHz and λ is the wavelength related to the frequency and

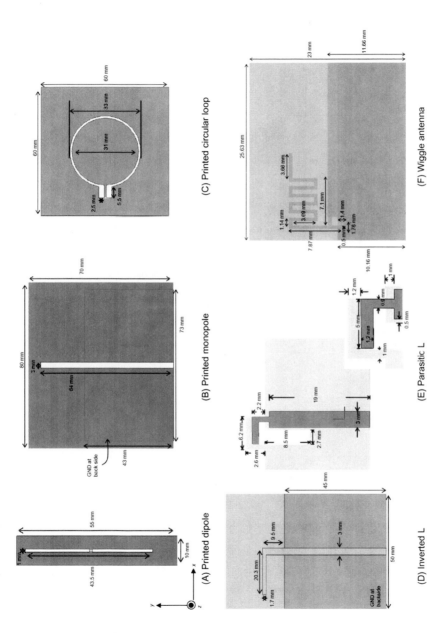

Figure 3.2 Schematics of antenna designs applied in the study: (A) dipole [12], (B) monopole [13, 14], (C) circular loop [12], (D) inverted L [15], (E) parasitic L [16], and (F) wiggle antenna [17]. All antennas are planar and printed on FR4 PCB board with dielectric constant of 4.6, loss of 0.002, and thickness h = 1.6 mm except for the wiggle antenna where h is reduced to 0.7 mm.

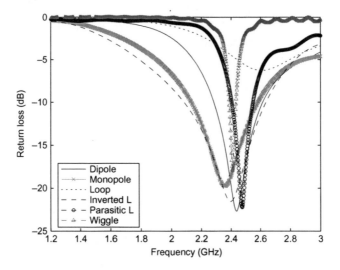

Figure 3.3 Return loss of the proposed antennas demonstrating excellent performance across the ISM band [10].

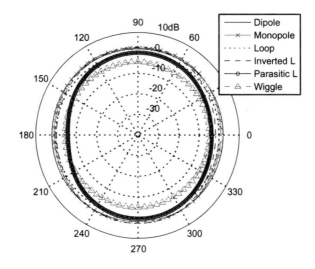

Figure 3.4 Radiation patterns of the applied antennas at 2.44 GHz in the azimuth plane (xz) [10].

speed of light), which is approximately 15.6 mm. The human tissue electric properties were defined at 2.4 GHz for all organs and tissues used including heart, lungs, muscle, fat, and skin.

The impedance matching performance of the detailed antennas is analyzed when antennas are placed at different locations on the body. Figure 3.6A presents the return loss of the printed dipole antenna

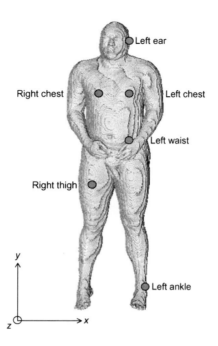

Figure 3.5 The male model used in the study, obtained from the Visible Male Project by the US air force with defined antenna positions applied in the investigation [5, 10].

when placed on the left side of the chest at various distances from the body. The results show detuning from free space resonance and that is due to changes in antenna effective length caused by the presence of the human lossy tissue with varying dielectric constant. The closer the antenna to the body the resonance frequency detunes the most, and again this is due to changes in effective dielectric media and also the electric length. The return loss results of printed monopole and parasitic L antennas (Figure 3.6B and 3.6C) shows slight detuning from free space resonance in comparison to dipole case and this is due to the presence of the ground plane which makes matching more stable and the percentage of substrate exposed to the lossy tissue is minimized. In parasitic L antenna case, the detuning from free space resonance is apparent; however, the effect of distance variation from the body is less apparent in contrast to dipole. The antenna narrow impedance bandwidth makes the small discrepancies between on-body and free space resonances of great importance in defining system reliability and efficiency.

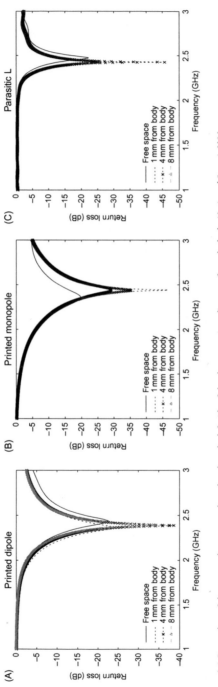

Figure 3.6 Return loss of the proposed antennas when placed on the left side of the chest at various distances from the body (1, 4, and 8 mm) [10].

Figure 3.7 shows the radiation patterns of the dipole antenna when placed on the left side of the chest and also on the left ear at 2.44 GHz. The patterns illustrate the loss caused by the presence of the human body, specifically the backward radiation, and also the increased directivity at specific angles due to the body curvature. Table 3.1 presents the body-worn antenna parameters of the various structures at different distances from the body. The antenna gain and efficiency for left side case is reduced in comparison to right chest and this is due to changes in human tissue physiological and electrical properties present at various parts of the body which directly influences the electric property and hence the power absorption. The further the antenna from the body, the higher the gain, and this is due to the increase in reflected power and reduction on absorbed energy.

Antennas with ground plane used for matching the antenna and also providing the necessary component to achieve the optimized performance, such as monopole and inverted L, proved to be least affected by varying distance from the body. On the other hand, printed dipole antenna resonance was detuned with different distance from the body, and the further the antenna from the body, the closer the resonance to free space operation. Inverted L, parasitic L, and wiggle antenna showed very narrowband performance for return loss and transmission loss analysis. This indicates that careful considerations need to be taken when applying these antennas for BANs or personal area networks since slight detuning introduces unreliable communication link. Angular radiation performance for all antennas provided stable performance across the band of interest (2.4−2.48 GHz) with main influence from the body curvature and size changes [10].

3.3 INTEGRATED ANTENNA IN COMPACT WIRELESS BODY-WORN SENSOR

For wireless sensor applications, antennas need to be efficient and immune from frequency and polarization detuning. Understanding the antenna radiation pattern for wireless sensors (when placed on the human body) is vital in determining the sensor performance. It is also important to specify how coupling into the propagation mode which may be a surface wave or free space wave or a combination of both occurs. If the sensor is placed too close to the human body, it will have low efficiency due to loss but good coupling to surface wave. On the

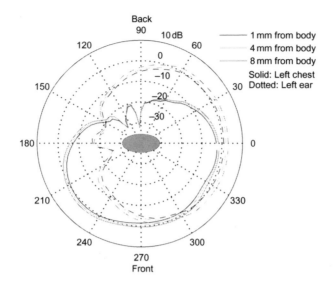

Figure 3.7 Radiation patterns of the printed dipole antenna placed on left side of the chest and the left ear at 2.44 GHz in the azimuth plane (xz).

Table 3.1 Comparison between Different Analyzed Antenna Parameters for Various Body-Worn Antenna Structures at Varying Distances from the Body

Antenna	Parameter/ Position	Right Chest	Left Chest	Left Ear	Left Waist	Right Thigh
				1/4/8 mm		
Printed dipole	Gain (dBi)	2.6/3.1/3.6	0.9/1.6/2.4	1.0/2.1/3.2	−/2.2/−	−/2.9/−
	Efficiency (%)	39/47/48	28/33/38	27/33/43	−/26/−	−/44/−
Printed monopole	Gain (dBi)	4.0/4.5/5.0	3.5/4.0/4.5	2.5/3.5/4.4	−/5.2/−	−/4.5/−
	Efficiency (%)	49/53/59	45/51/58	32/39/49	−/56/−	−/48/−
Circular loop	Gain (dBi)	2.0/2.6/3.0	1.4/−/3.0	1.7/2.3/3.3	−/5.3/−	−/3.7/−
	Efficiency (%)	30/34/38	25/−/35	26/30/36	−/49/−	−/39/−
Inverted L	Gain (dBi)	3.2/3.4/3.8	3.4/3.5/3.9	0.8/2.4/3.4	−/4.0/−	−/4.3/−
	Efficiency (%)	44/46/50	35/39/45	23/32/41	−/43/−	−/50/−
Parasitic L	Gain (dBi)	1.9/2.1/2.5	0.5/0.8/1.4	1/1.5/−	−/3.5/−	−/2.2/−
	Efficiency (%)	29/31/35	26/29/32	20/23/−	−/39/−	−/32/−
Wiggle antenna	Gain (dBi)	−4.2/−/−	−	−	−	−
	Efficiency (%)	21−1−	−	−	−	−

other hand, if it is placed too far from the body, the antenna efficiency will improve; however, coupling to surface wave will be poor. This is similar to classical very low frequency antenna over ground problem but parameters and antenna types will be very different. Hence it is essential to develop special antennas for wireless sensor networks that need monopole patterns for coupling to surface wave and patchlike patterns for surface/space wave links.

An example is used here to highlight the requirements and parameters involved in compact sensor designs from antenna prospective. This wireless sensor design is part of the development between Queen Mary, University of London [11], and Healthcare Devices and Instrumentation, Philips Research, The Netherlands (Figure 3.8), for operation in the unlicensed ISM band (2.4 GHz with ZigBee specifications). The sensor antenna design is restricted by many factors including the sensor size, chips placement, lumped element locations, and flexibility of the sensor structure to be shuffled with minimum cost and changes to antenna performance. The antenna can be compared to the circumference monopole, which is derived from the bent and inverted L antenna [18]. The sensor antenna performance is sensitive to lumped components, pins, and copper routings presence. The surrounding and adjacent components are modeled as a perfect conductor block around which the antenna is printed. The PCB board includes a ground plane and supply voltage copper sheets.

When the monopole is placed or printed on a dielectric material with permittivity other than 1, the antenna dimensions have to be modified in order to achieve performance at the frequency of interest. This leads to redefinition of antenna impedance by the approximation,

$$Z(\omega, \varepsilon_r) = \frac{1}{\sqrt{\varepsilon_r}} Z\left(\sqrt{\varepsilon_r}\omega, \varepsilon_0\right) \tag{3.11}$$

which leads to the antenna length being shortened by a factor $1/\sqrt{\varepsilon_r}$ to ensure the resonance is at the desired frequency. The metal pins and copper lines provide an extension to the antenna which increases the impedance value expected; however, due to the capacitive coupling between the antenna and the surrounding components, the antenna impedance tends to decrease compared to the general monopole impedance of 40 Ω [5]. Therefore, the radiation characteristics of the antenna are directly affected; hence changes in antenna gain and efficiency are expected.

Figure 3.8 Transceiver circuit layer of the medical sensor developed by Philips (applying Chipcon CC2420 transceiver chip from Texas Instrument [19]) with antenna presented as printed thin wire on the edges of the board: (A) top view, (B) bottom view, and (C) prototype fabricated sensor.

The printed circular monopole antenna within the sensor design is modeled with FR4 substrate ($\varepsilon_r = 4.6$ and thickness of 0.3 mm). The printed antenna thickness is 35 μm and the width of the line is 150 μm. The ground and supply voltage layers added have a diameter of 5.5 mm, thickness of 17.5 μm each, and separation between the layers of 80 μm. The actual antenna length is 31.5 mm. The antenna is a circumference quarter wavelength monopole. The complex impedance at the RF transceiver differential output is $115 + j180\ \Omega$; therefore, a matching circuit is applied in order to match the output to the single-ended monopole (matching to 50 Ω) [19]. Initial analysis of the monopole impedance at 2.4 GHz demonstrated the effect of adjacent components on antenna performance and hence antenna impedance in addition to the bend introduced in the antenna design. The antenna has a complex impedance of $35 + j320\ \Omega$ at 2.4 GHz. A capacitor of value 0.2 pF is introduced to match the antenna to the 50 Ω impedance seen at the matching circuit output. Figure 3.9 presents the sensor antenna return loss and the resulting narrowband caused by capacitor addition.

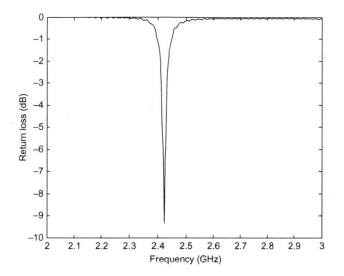

Figure 3.9 Simulated return loss of sensor antenna in free space with resonance around the desired frequency.

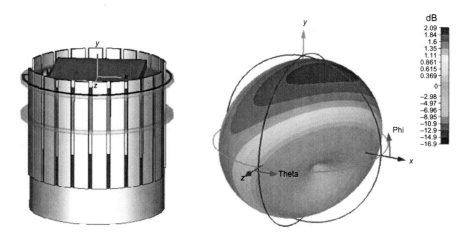

Figure 3.10 Sensor antenna radiation pattern numerically calculated at 2.4 GHz.

The 3D antenna radiation pattern at 2.4 GHz is shown in Figure 3.10. The pattern is different from that of a conventional vertical monopole due to introduced bend and also the effect of surrounding elements. The antenna gain calculated numerically is around 2 dB with efficiency of 77%. This demonstrates the potential reliable application of the sensor in free space with sufficient coverage area obtainable.

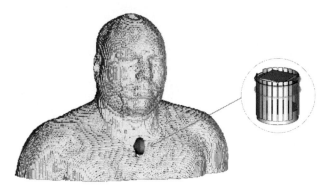

Figure 3.11 Sensor placed on the male model provided by the visible US human project.

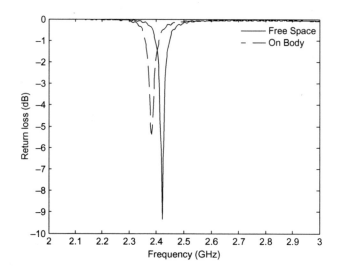

Figure 3.12 Return loss of sensor antenna when placed in free space compared to on-body return loss.

Body-worn sensor antenna performance is numerically investigated when sensor is placed on the human chest. Simplified human model which consists of a slab of lossy human is initially applied for cross referencing and evaluation. The tissue slab measures $120 \times 110 \times 44$ mm^3 with muscle electric properties ($\varepsilon = 52.8$ and $\sigma = 1.7$ S/m at 2.4 GHz). Once satisfactory initial results are obtained, the sensor is placed on the chest of the detailed multilayer human model, namely the visible male model developed by the US air force (http://www.brooks.af.mil/AFRL/ HED/hedr/), as shown in Figure 3.11.

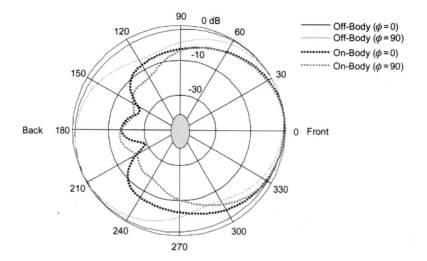

Figure 3.13 Azimuth plane radiation pattern of sensor antenna when placed in free space and on the body.

The simulated return loss of the sensor shows the slight detuning due to the presence of lossy tissues with sensor placed 2 mm away from the body as demonstrated in Figure 3.12. Although the detuning effect is minimum, the narrowband nature of the antenna (with capacitor used for matching) signifies such changes and its direct influence on radiated energy.

The antenna gain when placed on the body is increased to 2.4 dB caused by reflections from the human body, which at this high frequency is considered as a large reflector due to high losses and also very small penetration depth. The azimuth plane radiation patterns shown in Figure 3.13 illustrate the effect of human body on the antenna performance and the reduced radiated power in the backward direction with front to back ratio of around −30 dB.

CHAPTER 4

Cooperative and Low-Power Wireless Sensor Networks for Body-Centric Applications

Wireless body area networks (WBANs) are attractive solutions that can be used in healthcare, sport, and lifestyle monitoring applications, enabling constant screening of health data and constant access to patients regardless of their current location or activity and with a fraction of the cost of a regular face-to-face examination. Devices like ECGs, pulse oximeter, blood pressure, insulin pumps, and blood glucose can be coupled with wireless and wearable communication sensors. Particularly, a 3-, 4-, 5-, or 10-leads ECG monitor system can be conveniently integrated into a sensor network, with heart activity data captured and monitored by wearable circuitry (on-body segment) and wirelessly transmitted to a nearby listening device (off-body segment). To successfully deploy a WBAN performing long term and continuous healthcare monitoring, it is critical that the wearable devices are small and lightweight, lest they be too intrusive on patient lifestyle.

Research on sensor network has been carried out using small, low-power digital radio based on an IEEE 802.15 standard [20], a high-level communication protocol suitable for WBANs [21, 22]. The most straightforward approach to deploy a WBAN is considering single-hop (SH) communications between the sensors and the sink. However, the body impact on the signal can result in path losses usually larger than 50 dB [23]. Due to these high losses, direct communication between the sensors and the sink will not always be convenient (or even possible), especially when extended sensor lifetime is targeted for ultra-low range transceivers [24−27]. Multi-hop (MH) communications exploit spatial diversity as an advantageous solution (and sometimes even as an absolute requirement) to ensure connectivity, improve link reliability, and possibly improve the energy efficiency of WBANs [28].

In a relay MH network, each sensor is required to transmit or relay information packets, while in cooperative MH network, each sensor can perform sequentially both operations. An example of relay WBAN benefits is introduced in [29], where spatial diversity gain is analyzed

Co-operative and Energy Efficient Body Area and Wireless Sensor Networks for Healthcare Applications.
DOI: http://dx.doi.org/10.1016/B978-0-12-800736-5.00008-8

for a two-relay assisted transmission link. Tree cross-layer protocols such as CICADA [30] and WASP scheme [31] aimed to achieve WBANs reliability and low delay, although no considerable attention is focused on balancing the power consumption between the interconnected sensors [32]. Several researchers also attempted to design energy-aware MH protocols, considering also different metrics such as delay and reliability as quality of service requirements [33–36]. Although these studies demonstrate that MH communications are suitable to overcome link blockage in sensor WBANs, the energy efficiency compared to the SH schemes is still an open issue and depends on several system parameters, including chipset implementation, sensors distance, and network topologies.

The main objective here is to compare with real equipment the performances of SH and MH cooperative schemes for a WBAN. To assess the trade-off between these important parameters, real-life measurement scenarios are taken with IEEE 802.15.4/Zigbee compliant radio sensors, operating at the ISM 2.45 GHz frequency band. The power margins, the data throughput r, the packet delivery ratio, and the average energy consumption are selected as main performance criteria. The sensors generate and transmit data at regular intervals with a throughput suitable for ECG constant monitoring system. The MH cooperation is the scheme of choice for this study, as the deployment of dedicated relays would unnecessarily increase the total number of sensors on body.

4.1 PRACTICAL CONSIDERATIONS OF THE BODY-CENTRIC WIRELESS SENSOR NETWORK

A prototype synchronous sensor network at 2.4 GHz is set up, where each sensor consists of a Sentilla Perk mote [39], compliant with the IEEE 802.15.4/Zigbee standard. A total number of four (with index $i = 1, 2, 3, 4$) sensors were placed on a human volunteer (each attached on head, left leg, left wrist, back) in sitting posture as shown in Figure 4.1, where a sensor acting as sink is placed in the waist area. This is a representative scenario for patients who are resting for a major part of the day. Sensor 4 was placed on the volunteer's back diametrically opposite to sensor 3. The sensors are placed such that batteries are closest to skin, with the antennas being further away. With respect to the sink, two sensors are in line of sight (e.g., 1 and 2) while

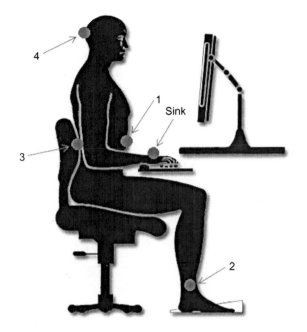

Figure 4.1 Displacement map of four sensors and a sink network on volunteer body.

two others are in non-line of sight (e.g., 3 and 4). Experiments were run on an office indoor scenario.

The embedded CC2420 RF transceiver provides eight transmission levels ranging from −25 to 0 dBm power output. All experiments correspond to multi-point-to-point routing in which data from all other sensors are sent to sensor 4 which is designated as the on-body sink sensor (see Figure 4.1). The sink collects raw data and sends statistics to an off-body server using a wireless link. The network operations can ideally be cyclically repeated and they can be divided in three main phases: (1) setting-up of the routing tree topology, (2) time slot transmission synchronization, and (3) data transmission. The first two phases can be ranked as start-up phases and the latter as steady-state phase. The sensors send routing messages in phase 1, dummy messages in phase 2 for synchronization purposes, and actual data messages during phase 3. The network performs cycles of the three phases with periodicity T_N to adapt its topology to the body movements and postural and environment changes. After preliminary tests, a time-synchronous architecture approach was selected as the best suited to maximize the data delivery ratio.

4.1.1 Topology Update (Phase 1)

Two routing protocols can be used: a basic SH protocol to provide a baseline, and the minimum cost forwarding network routing protocol proposed by Fan et al. [40]. This power-efficient routing algorithm has been implemented on the top of the standard connection functionality provided by the Sentilla motes kit. From previous published works [41], the Received Signal Strength Indicator (RSSI) seems to provide a good estimation of packet loss rates; for example, RSSI of -90 dBm or larger always corresponds to a packet received ratio (PRR) of 95% or more. The RSSI comes from the CC2420 built-in register, whose values are estimated in accordance to [38]. The RSSI register value $RSSI_{VAL}$ can be referred to the power P_{RF} at the RF pins by using the following equations:

$$P_{RF} = RSSI_{VAL} + RSSI_{OFFSET} \qquad (4.1)$$

where the $RSSI_{OFFSET}$ is approximately -45 dB (e.g., if reading a value of -20 from the $RSSI_{VAL}$ register, the P_{RF} is approximately -65 dBm). The $RSSI_{VAL}$ can directly be related to the path loss L_p and to the transmit power P_{Tx} according to

$$RSSI_{VAL} = P_{Tx} - L_P - RSSI_{OFFSET} \qquad (4.2)$$

As transmit and receiving antenna gain cannot be explicitly estimated because of the relative orientation and body impact, they are considered as embedded in the L_P term. The routing algorithm adopted seeks to achieve minimum cost from each sensor toward the sink, where costs are proportional to the RSSI. In SH case, each sensor transmits by default to the sink so no routing data is required, while in MH protocol case, each sensor retains the next hop target sensor address to build the tree topology.

4.1.2 Synchronization (Phase 2)

The transmit time slots are synchronized in phase 2 using beacons with a unique sensor address periodically sent by the sink. In this phase, each sensor is constant in receiving mode, listening for sink beacons and other sensor messages. If a beacon is received, the sensor sends a packet to the next hop target sensor (or to the sink in case of SH scheme). If a packet is received in MH scheme, it is simply relayed to the next hop. As the sink knows the total number of sensors (but neither the network topology nor the latency), it does not send a new beacon until a packet is received from the target sensor. As the message delay is different per each sensor, the total network throughput is expected to be lower in MH scheme. Similar to the sink, the sensors can synchronize their wake-up time for receiving the packets from neighborhood sensors. After phase 2, the sink and the sensor have set a wake-up time, and no beacons are required anymore.

4.1.3 Transmission (Phase 3)

At the end of phase 2, the communication between sensors is time slotted according to the wake-up times to avoid idle listening and save power. Each sensor regularly transmits data packets of 75 bytes of payload. In case of MH protocol, a sensor is capable of relaying a data packet from neighborhood sensors immediately after reception, with no data buffering. The sensor operation type (e.g., transmit or transmit and relay) of each sensor depends on the network topology and it can dynamically change every cycle T_N. Considering a sensor in transmit operation type as shown in Figure 4.2, the tasks are divided into three time slots: in T_P, the sensor generates data to transmit, in T_{Tx}, the sensor transmits the data packets, and in T_S, the sensor is in sleeping mode. T_{Tx} is fixed and empirically estimated to be ~ 108 ms. This value data serialization, a method of transforming Java objects into a byte stream (binary form), can be sent and received over the radio link. Thus, the actual time required to transmit data is <100 ms. The sensor sleeping time is T_S varying according to the synchronizations, while the active time is defined as $T_A = T_{Tx} + T_P$, where T_P is ~ 72 ms.

In case of relay operations, the sensor is on receiving mode for a maximum time of T_W, forwards the packet to the next sensor during T_{Tx}, and sets itself back in sleeping mode. The receiving mode includes the two listening and data receiving tasks. As some drifts in synchronization can occur and T_W can expire before the transmission is successfully completed,

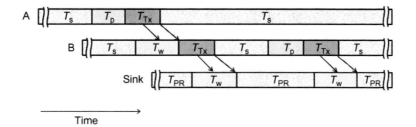

Time

Figure 4.2 Packet data communication tasks for a (A) transmit sensor, (B) relay sensor and sink sensors operation modes (blocks are not in scale).

$T_W > T_{Tx}$ is chosen to introduce a safety time margin. The system throughput depends also on the sensor speed to process the data packets within T_W. A sensor sink with a faster processor would possibly allow a smaller T_W. Finally, the sink switches between two sequential tasks, namely setting itself at scheduled time on receiving mode for a maximum of T_W and performing the received data processing during T_{PR}. The path losses are estimated during the phase 1, and there are no means to refresh these values until the following topology update. Moreover, the topology updates cannot be as frequent as the link variations.

Although the alignment conditions of the antennas might be different for sensors' individual positions, it is not possible to de-embed from the L_P term the antenna gains as the antenna characteristics are highly affected by human body. The margin M can be linked to the channel statistics considering the path loss L_P as a random variable given by $L_P = L_{AVG} + L_S$, where L_{AVG} is the value predicted by the average path loss model, and the shadowing component L_S is a zero-mean Gaussian random variable with standard deviation σ. In order to provide communications with a given percentage reliability, an extra margin $L_S = t \cdot \sigma$ has to be added. The value of t can be calculated according to

$$t = \sqrt{2} \cdot erfc^{-1}[2 \cdot (1 - p)] \tag{4.3}$$

where $erfc^{-1}$ is the inverse of the standard cumulative error function, and p is the percentage of reliability that is required [28]. In Ref [37], a model is presented where the sensor decrements the available energy according to the following parameters: (a) the specific transceiver characteristics, (b) size of the packets, and (c) the bandwidth used. The following equations represent the energy used for each information bit when a data packet is transmitted Eq. (4.4) or received Eq. (4.5),

$$\tilde{E}_{Tx} = V \cdot I_{Tx}(P_T) \cdot p \tag{4.4}$$
$$\tilde{E}_{rx} = V \cdot I_{rx} \cdot p \tag{4.5}$$

where V is the transceiver typical supply voltage, p is the bitrate, and $I_{Tx}(P_T)$ and I_{rx} are the transmit and receive currents, respectively. The transmit current is a nonlinear function of the transmit power P_T, and a third-order polynomial interpolation has been used to model data from the CC2420 chipset datasheet [38]. Equation (4.6) shows the interpolating functions:

$$I_{Tx}(P_T) = 4 \cdot 10^{-4} \cdot P_T^3 + 3 \cdot 10^{-2} \cdot P_T^2 + 0.9 \cdot P_T + 17.4 \tag{4.6}$$

where P_T is expressed in dBm and the current in mA. The ability to control the transmission power is available on the CC2420 radio platforms, providing eight transmission levels (ranging from -25 to 0 dBm output) selectable at run-time by configuring a register. This would correspond to eight values of transmit power P_T. The transmit currents would depend on the margin required at each link, while the receive current is constant. Thus, to transmit k bits with a given transmit power P_T, the radio expends $E_{Tx} = \tilde{E}_{Tx} \cdot k$, while to receive this message, the radio expends $E_{rx} = \tilde{E}_{rx} \cdot k$. Equation (4.4) can be used to describe the SH sensor energy expended in transmitting a SH k-bit message from a sensor to the sink for a given P_T. In case of MH, each sensor sends a message to the next hop sensor on the way to the sink. Thus, the relay sensor would require receiving, retransmitting data from neighbor sensors, and transmitting its own data. The total energy consumed would be:

$$E_{tot} = [\tilde{E}_{Tx} + \tilde{E}_{rx}] \cdot k \cdot n + \tilde{E}_{Tx} \cdot k \qquad (4.7)$$

where n is the number of neighbor sensors. It is assumed that all sensors transmit the same message size and the received message has the same size of the transmitted message. The tree topology and the resulting higher forwarding overhead make the sensors near the sink perform worse than those further away. For a given path loss L_P, M is proportional to P_T. As each sensor in this study transmits its own ECG biomedical data, the network lifetime definition is based on the first sensor failing time (e.g., running out of energy). We borrow the analytical model used in [42] to calculate the sensors and the network lifetime. Each sensor is assumed to have an initial battery energy B_i. Then, the lifetime of a sensor i is given by

$$T_i = \frac{B_i}{\{[\tilde{E}_{Tx} + \tilde{E}_{rx}] \cdot n + \tilde{E}_{Tx}\} \cdot r} \qquad (4.8)$$

We define the network lifetime T_{net} to be the time until the first sensor runs out of energy, that is,

$$T_{net} = \min\{T_i\}, \quad \text{for } i \in N \qquad (4.9)$$

4.2 EXPERIMENTAL INVESTIGATIONS AND ANALYSES

Considering a static posture of the patient, we assume a constant time average for RSSI per each link. Each body link has been preliminarily characterized in terms of measured average RSSI. Data are stored in the device's memory and associated to each link before the experiments. As the links costs are now fixed, a single network cycle T_N is enough for each test. This approach has the benefits of comparing the SH and MH energy consumptions for the same network topology, enabling a separate study of the packet losses due to the synchronization drifts from those where the P_{RF} drop below the sensitivity threshold and the repeatability of the results. Once the networks in terms of PRR and r are measured, the sensor energy consumption is modeled offline and studied as function of M.

4.2.1 Network Topologies and Body Links Characterization

A preliminary characterization of the network topology in terms of link cost and time variability is performed. Per each link (e.g., γ_{13}), a data packet was sent every 1 s at 0 dBm of transmit power, for an observation time of 2 min. Each measurement was repeated three times and data were merged in a single history vector for each link. Per each packet, an RSSI measurement based on the Zigbee standard was stored and the path losses statistics are derived from these values. While taking measurements, the volunteer was allowed to perform changes in the posture, as naturally happens in such scenario. Figure 4.3 shows the averaged RSSI and received power from measurements of the sitting postural setup. Higher RSSI values correspond to lower link costs. In case of SH scheme, the sensors 1, 2, 3, and 4 can only transmit directly to the sink.

In case of MH scheme, the routing protocol sets the sensor 1, 2, and 4 to communicate directly to the sink, as these links have a lower link cost if compared with any other MH link combination. The sensor

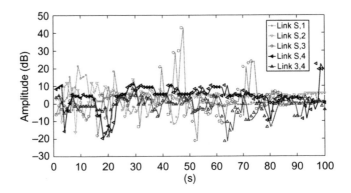

Figure 4.3 Averaged RSSI and RF powers in dBm for sitting postural setup.

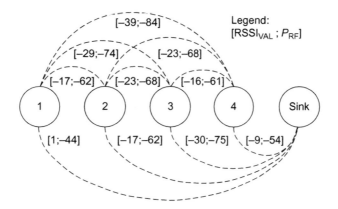

Figure 4.4 Sample L_S history during the first 100 measurement seconds with sampling rate of 1 s.

Table 4.1 Average Path Loss, Available Margins, and Statistical Parameters of L_S Per Each Link of Interest						
Link	L_{AVG} (dB)	M_0 (dB)	L_S (dB)			
			σ	Range	Min.	Max.
γ_{S1}	−44	24	7.2	53	−31.4	21.52
γ_{S2}	−62	42	6.7	43	−25.11	17.89
γ_{S3}	−75	20	8.1	66	−23.02	42.98
γ_{S4}	−54	34	7.4	44	−21.57	22.43
γ_{34}	−61	41	5.0	32	−21.33	10.67

3 transmits to the sensor 4 and the latter acts as relay. In fact, considering the RSSI, the γ_{S3} link cost (where S stands for sink) is higher than the sum of γ_{34} and γ_{4S} link costs (e.g., $(-16) + (-9) < -30$). This can potentially result in transmit power margin for the sensor 3 of 14 dB if compared to the SH case. The P_{RF} values are not numerically suitable as link costs for the routing algorithm. In fact, the sum of any MH links combination will not be lower than the SH links, even for MH power-wise convenient routes. As discussed before, only the four links relative to the sink are of interest for both SH and MH, while γ_{34} is of interest for MH only. The L_S time histories of these links are shown in Figure 4.4, while Table 4.1 shows the statistical parameters. The standard variation σ spans from 5 to 8.1 dB, while the power range is up to 66 dB.

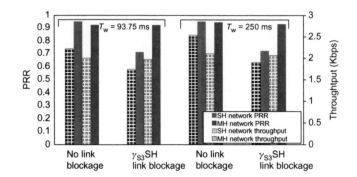

Figure 4.5 Total packet delivery and throughput r for the SH and MH cases.

4.2.2 Network Packet Delivery Ratio and Data Flow Rate Performances

From Table 4.1, the sensor 3 has M_0 of 20 and 34 dB for the SH and MH cases, respectively. Moreover, σ is 8.1 and 5.0 dB for the γ_{S3} and γ_{34} cases, respectively. According to formula (4.3), the probability of exceeding M_0 (or link blockage probability) is about $7 \cdot 10^{-3}$ for the SH and about $5 \cdot 10^{-12}$ for the MH case. Thus, a γ_{S3} blockage is significantly more likely than a γ_{34} link blockage. Figure 4.5 shows the SH PRR (primary y-axis on the right) and r (secondary y-axis on the left) for γ_{S3} with and without link blockage. The results are compared with the MH PRR and r with no link blockage on the γ_{34} link. The cases for $T_w = 93.75$ and 250 ms are considered. The PRR is >0.9 for both SH and MH schemes with no links blockage. In case of MH network, it shows the capability of sensor 4 to receive and route to the sink at least the data from the sensor 3 with a PRR comparable (e.g., PRR > 0.9) to the SH case with no blockage. In case of γ_{S3} link blockage, the SH PRR degrades about 23% compared to the MH with no link blockage for both T_W cases, while the SH r reduction is 21% and 25% for the T_W case of 93.75 and 250 ms, respectively. This means that the MH topology can be used to overcome SH link blockages theoretically without PRR degradation.

The PRR (and consequently r) depends directly on the T_W value. For this reason, both the PRR and r are preliminarily studied against the T_W to maximize the data r while keeping at minimum the packet losses. Figure 4.6 compares the measured PRR (primary y-axis on the right) and the r (secondary y-axis on the left) against T_W for SH and MH schemes with the link costs as defined in Table 4.1. The average

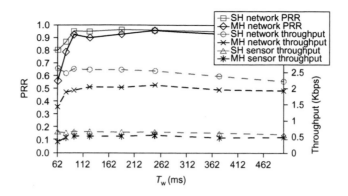

Figure 4.6 PRR, total, and average r against T_W for SH and MH schemes.

per each sensor and the total network r are included. For $T_W \geq 93.75$ ms, the PRR is ≥ 0.9 for both schemes.

Within this range, the MH scheme exhibits a PRR reduced by about 0.03 on average when compared to SH. This can be explained with the lower robustness to synchronization drifts of the MH branch, where the combined probability for the sink and relay sensor 4 waiting time T_W to expire before the data packet transmission is completed are doubled if compared to any sensor-link direct link. The PRR becomes even lower for $T_W < 93.75$ ms. Considering an ideal packet data communication, with time slotted synchronization mechanism, a PRR = 0.95 and $T_A \approx 180$ ms (e.g., with a transmission duty cycle $\delta \approx 0.18$), $T_{Tx} = T_W$, the maximum r per each sensor is 0.7912 and 0.63 Kbps for SH and MH schemes, respectively. The maximum r achieved in the real-world platform is 0.7 and 0.5 Kbps for $T_W = 125$ ms for SH and MH schemes, respectively.

Assuming an ECG signal sampled at 360 Hz and a resolution of 11 bit/sample, the resulting signal data rate is 3.96 Kbps. However, well-known compression techniques based on wavelet (as shown in [43]) can lower the ECG data rate down to 0.33 Kbps, making it fully compatible with data rate supported by this network. The transmission of the MH branch (e.g., γ_{S43}) requires twice the time for SH branch transmission (e.g., γ_{S3}), lowering the average sensors r of about 20%. The network r (or r at the sink input) are 2.6 and 2.1 Kbps for $T_W = 125$ ms of SH and MH schemes, respectively. It is envisaged that a more efficient synchronization algorithm would increase considerably the MH r. For instance, the sink is idle while the sensor 3 is transmitting to 4; however, it could receive data from sensor 2 as well to provide cooperation. Under these conditions, SH and MH schemes can potentially have the same r.

4.2.3 Analytic Energy Consumption Analysis

As path loss L_S can change significantly in time even for sitting posture because of body movements, it is important to compare M against the energy consumption. The relationship between the transmit power and the sensor current consumption is not linear, and it can be retrieved from the formula (4.8). The energy consumption can be estimated as a function of data packets transmitted, based on the exact circuitry being used. As a 1.8 V chipset voltage is used and the defined bitrate is 250 Kbps [39], $\tilde{E}_{rx}$ is 135.4 nJ/bit, while $\tilde{E}_{Tx}$ at $P_T = 0$ dBm is 125.28 nJ/bit. The microcontroller energy costs are not considered. From Ref. [39], the current consumption in idle and sleep modes are significantly smaller (>500 μA) compared to the maximum transmit and receive currents (17.7 and 18.8 mA, respectively) and they are not considered neither.

The energy consumptions estimation only represents the energy per bit dissipated in the transceiver. As the extra energy dissipated during overhead processing (data generation, data serializations, etc.) and the media access control related (such as the waiting time T_W) are not considered, this approach provides more general results as the energy is approximated using only the network topology, the transmitted power, the chipset implementation, the bitrates, and the target reliability. Equations (4.4) and (4.5) are then used to estimate the energy to receive and transmit the 75 bytes message payload with power value of P_T. The energy consumption at sensor 4 is estimated using formula (4.7) and with $n = 1$. In MH case, the sensor 4 receives and retransmits the 75 bytes sent from sensor 3.

Figure 4.7 shows a comparison of E_{tot} against the power margins per each sensor. As the sensors 1 and 2 have the same energy consumption for both schemes, a comparison between sensors 3 and 4 is representative for the two schemes. Considering a power margin M of 5 dB, for instance, E_{tot} for the sensor 4 is 155.4 μJ while E_{tot} for the sensor 3 is 48.89 μJ. As the sensor receiver energy (e.g., 81.2 μJ) is higher than the maximum transmit energy (e.g., 75.1 μJ), the minimum

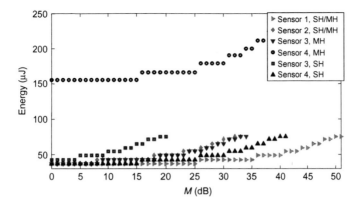

Figure 4.7 Total energy consumptions against power margins for SH and MH networks.

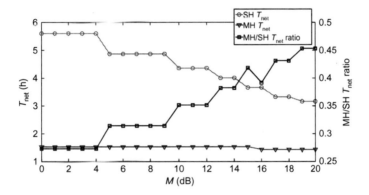

Figure 4.8 T_{net} for SH and MH compared against the power margin M for $\mathfrak{r} = 0.7$ Kbps.

E_{tot} of the sensor 4 (e.g., 155.4 µJ) will be always higher than the maximum E_{tot} for the sensor 3 or 4 (e.g., 75.1 µJ).

Considering now a statistical characterization of the path loss using formula (4.3), M can be derived for a communication with a reliability p.

4.2.4 Network Lifetime
Implementing Eqs. (4.8) and (4.9), Figure 4.8 derives the T_{net} for SH and MH schemes for an M ranging from 0 to 20 dB (corresponding to the maximum M_0 for the sensor 3 in the SH scheme). Note that the T_{net} corresponds to the lifetime of sensors 3 and 4 for SH and MH schemes, respectively (as they have the shortest lifetime for all the M range). The initial total energy of each sensor is set to be $B_i = B_j = 1$ to be consistent

with previous works [44]. The energy per unit of bit information values for receiving and transmitting at $P_T = 0$ dBm (e.g., E_{rx} and E_{Tx}) are as given previously, and it is assumed that $n = 1$ and $r = 0.7$ Kbps. Although the measured r is different for the SH and MH schemes, the same value is considered in both cases for consistency. The SH scheme lifetime spans from 5.6 to 3.2 h according the M, while the MH one spans only from 1.5 to 1.4 h. Assuming the same M for both sensors, the network lifetime of the MH network ranges from 27% to 45% of the SH lifetime. As $M = 18.8$ dB provides a $T_{net}^{99\%} = 3.1$ h for SH network, and $M = 17.2$ dB provides a $T_{net}^{99\%} = 1.6$ h for the SH, the MH lifetime corresponds to 51% of SH one.

CHAPTER 5

Design of Body-Worn Radar-Based Sensors for Vital Sign Monitoring

In this section, we look into the design of a wearable vital sign sensor for future cooperative, energy-efficient body area network (BANs). An ideal solution is to implement a body-worn sensing system that does not require any contact to the user. One's vital health parameters could be continuously monitored, and wireless transmission of these data could allow remote signal analysis. In the previous works, the use of ultra-wideband (UWB) radar has been suggested for unobtrusive monitoring of patient's vital signs [45, 46]. Due to the high resolution of UWB signals, the expansion of the chest cavity will cause periodical variations to the received signal, which can be exploited to estimate the respiration rate. In this section, the human chest is modeled as a planar multilayered medium with one-dimensional inhomogeneity, where the electromagnetic property is piecewise constant in each layer. Subsequently, a multiray propagation model is proposed to take into account significant multipath components contributed to the received signal including the line-of-sight (LOS) component, and the transmitted signals reflected off the air/skin/fat/muscle interfaces. Based on the proposed model, the time-of-arrival information of the first echo is utilized to measure the breathing frequency. Subsequently, we perform simulation studies to validate the derived model and to analyze the system performance.

5.1 SYSTEM ARCHITECTURE AND PROPAGATION MODEL

Let us consider a generic breath monitor as shown in Figure 5.1A. Both the transmitter (Tx) and receiver (Rx) are attached at two ends of a physical holder, which is worn on clothes or blankets without direct skin contact. The wearable holder is placed near the chest wall in order to accurately detect respiratory movements. In recent years, low-cost, low-profile textile UWB antenna for BANs has been successfully deployed [47], which would facilitate such an application. When the UWB pulse is transmitted from the Tx, the signal will be echoed

Co-operative and Energy Efficient Body Area and Wireless Sensor Networks for Healthcare Applications.
DOI: http://dx.doi.org/10.1016/B978-0-12-800736-5.00009-X

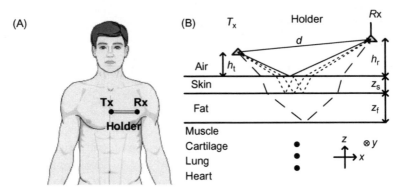

Figure 5.1 (A) Placement of the antennas close to the human chest wall and (B) the multiray propagation model.

back to the Rx from the interfaces among different layers of living tissues as illustrated in Figure 5.1B [46]. Six tissue layers in the thorax have been identified: skin, fat, muscle, cartilage, lung, and heart. As the intrinsic impedances of muscle, cartilage, lung, and heart are nearly the same [46], they will be treated as one homogeneous medium and referred to as "muscle" in the subsequent discussion. Without loss of generality, we consider the two-dimensional propagation scenario and the perpendicular polarization or *TE wave* where the electric field is transverse to the direction of propagation. By approximating the propagation waves as geometrical optic rays and taking into account the multiple reflections and transmissions in different regions of the multilayer medium (see also Figure 5.1B), the electric field and path delay time of various rays received at the Rx can be readily calculated.

The UWB waveform used is the fifth derivative of a Gaussian pulse, which is given by

$$u(t) = \left(-\frac{t^5}{\sqrt{2\pi}\sigma_t^{11}} + \frac{10t^3}{\sqrt{2\pi}\sigma_t^9} - \frac{15t}{\sqrt{2\pi}\sigma_t^7} \right) \exp\left(-\frac{t^2}{2\sigma_t^2} \right) \quad (5.1)$$

where $\sigma_t = 101 \times 10^{-12}$ s. The multiple cole–cole dispersion model can be used to describe the variation of dielectric properties of tissues as a function of frequency [48]:

$$\hat{\varepsilon}(\omega) = \varepsilon_\infty + \sum_m \frac{\Delta\varepsilon_m}{1 + (j\omega\tau_m)^{1-\alpha_m}} + \frac{\sigma}{j\omega\varepsilon_0}, \quad m = 1, 2, \ldots \quad (5.2)$$

where $\hat{\varepsilon}$ is the complex relative permittivity and ω is the angular frequency. The numerical values of model parameters in Eq. (5.2) for various tissue types are listed in Table 5.1 [48]. Subsequently, the backscatter signals can be obtained by performing an inverse discrete Fourier transform for a set of frequencies spanning the whole UWB spectrum. The following parameters are applied in the numerical example (see also Figure 5.1B for the definitions of variables): $h_t = h_r = 8$ mm, $d = 8$ cm, $z_s = 1$ mm, and $z_f = 10$ mm. The backscatter signal characteristics are shown in Figure 5.2. The indexes 1−8 along the x-axis correspond to the following pulse echoes: LOS path (1); signal reflected from the air−skin interface (2); signals reflected from the skin−fat interface with $N = 1$ (3) and $N = 2$ (4); signals reflected from the fat−muscle interface with $M = 1$, $N = 1$ (5), $M = 1$, $N = 2$ (6), $M = 2$, $N = 1$ (7), and $M = 2$, $N = 2$ (8). M and N denote the numbers of reflections on the fat-to-muscle and skin-to-fat interfaces, respectively. As can be seen from Figure 5.2A, the most dominant pulse echo is the signal reflected from the air−skin boundary (corresponding to index 2), which is about 25 dB larger than the second strongest wave (corresponding to index 3).

Since the echo from the air-to-skin boundary is strongly overwhelming the echoes from other boundaries, the multiray model can be simplified into a two-ray formulation:

$$r(t) = \chi_0\mu_0(t - \tau_0) + \chi_1\mu_1(t - \tau_1) \approx \chi_0\mu_0\left(t - \frac{d}{c}\right) + \chi_1\mu_1\left(t - \frac{d + 2h_t h_r/d}{c}\right)$$

(5.3)

where χ_0 and χ_1 are the amplitudes of the direct path and backscatter signal from the air-to-skin boundary, respectively. μ_0 and μ_1 are the corresponding pulses, and τ_0 and τ_1 are the time delays. c is the speed of electromagnetic wave in the air. The approximation in Eq. (5.3) is valid when $d \gg h_t$, h_r. In real-life applications, repetitive UWB pulses are transmitted to probe the chest cavity. The pulse repetition time T_s should be sufficiently shorter than the breathing period. Effectively, the received waveforms are recorded at discrete instants $t = nT_s$ ($n = 1, 2, \ldots, Q$). Moreover, due to the periodical respiratory activity, the heights of the Tx and Rx antennas, h_t and h_r, vary periodically about nominal heights $\overline{h}_t$ and $\overline{h}_r$:

$$h_t(n) = \overline{h}_t + \sum_{k=1}^{\infty} A_{t,k}\sin(2\pi kf_0 nT_s + \varphi_{t,k})$$

(5.4)

Table 5.1 Parameters Used to Predict the Dielectric Properties of Tissues

Tissue type	ε_∞	$\Delta\varepsilon_1$	τ_1 (ps)	α_1	$\Delta\varepsilon_2$	τ_2 (ns)	α_2	$\Delta\varepsilon_3$	τ_3 (μs)	α_3	$\Delta\varepsilon_4$	τ_4 (ms)	α_4	σ
Skin (dry)	4	32	7.23	0	1100	32.48	0.2	0			0			0.0002
Fat (not infiltrated)	2.5	3	7.96	0.2	15	15.92	0.1	3.3×10^4	159.15	0.05	1×10^7	7.958	0.01	0.0100
Muscle	4	50	7.23	0.1	7000	353.68	0.1	1.2×10^8	318.31	0.1	2.5×10^7	2.274	0	0.2000

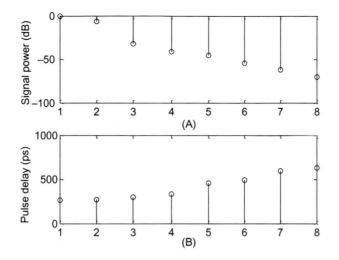

Figure 5.2 Backscatter signal characteristics: (A) signal power and (B) pulse delay.

$$h_r(n) = \overline{h}_r + \sum_{k=1}^{\infty} A_{r,k}\sin(2\pi k f_0 n T_s + \varphi_{r,k}) \qquad (5.5)$$

where f_0 is the respiration rate, $A_{t,k}$ and $A_{r,k}$ are the coefficients of the Fourier series expansion of the periodical functions h_t and h_r, respectively. $\varphi_{t,k}$ and $\varphi_{r,k}$ are the phases. The pulse delay time for echo at the air-to-skin boundary is then computed by substituting Eqs. (5.4) and (5.5) into Eq. (5.3):

$$\tau_1(n) = \frac{d^2 + 2h_t(n)h_r(n)}{cd}$$

$$= \frac{1}{cd}\{d^2 + 2\overline{h}_t\overline{h}_r + \sum_{k=1}^{\infty} 2\overline{h}_r A_{t,k}\sin(2\pi k f_0 n T_s + \varphi_{t,k})$$

$$+ \sum_{k=1}^{\infty} 2\overline{h}_t A_{r,k}\sin(2\pi k f_0 n T_s + \varphi_{r,k})$$

$$- \sum_{k_t=1}^{\infty}\sum_{k_r=1}^{\infty} A_{t,k_t} A_{r,k_r}[\cos(2\pi(k_t + k_r)f_0 n T_s + \varphi_{t,k} + \varphi_{r,k})$$

$$- \cos(2\pi(k_t - k_r)f_0 n T_s + \varphi_{t,k} - \varphi_{r,k})]\} \qquad (5.6)$$

As can be seen from Eq. (5.6), the arrival time of the first echo consists of a series of breathing-frequency harmonics. To provide intriguing insight into the critical design parameters, we consider a simplified scenario when the antenna heights are sinusoidal. In such case, Eq. (5.6) can be simplified as,

$$\tau_1(n) = \frac{d^2 + 2\overline{h}_t\overline{h}_r + A_tA_r\cos(\varphi_t - \varphi_r)}{cd}$$

$$+ \frac{2\sqrt{\overline{h}_t^2 A_r^2 + \overline{h}_r^2 A_t^2 + 2\overline{h}_t\overline{h}_rA_rA_t\cos(\varphi_t - \varphi_r)}}{cd}\sin(2\pi f_0 nT_s + \psi)$$

$$- \frac{A_tA_r}{cd}\cos(4\pi f_0 nT_s + \varphi_t + \varphi_r)$$

$$(5.7)$$

where the subscript k has been omitted for the sake of clarity and ψ is a deterministic phase. Several important observations can be made from Eq. (5.7). First, the spectrum of the first-echo delay time comprises the fundamental component f_0 and the second harmonic $2f_0$. Second, when the periodical functions h_t and h_r are in phase ($\varphi_t = \varphi_r$), the maximum amplitude of the fundamental component is obtained. Furthermore, this value is always larger than the amplitude of the second harmonic because $\overline{h}_t \geq A_t$ and $\overline{h}_r \geq A_r$ for nonzero antenna heights. It is worth mentioning here that this would be the most common situation in real-life deployment, since the chest-cavity movement will increase or decrease the antenna heights simultaneously. Finally, a smaller LOS distance yields larger amplitudes of harmonics, thereby leading to more accurate frequency estimation.

5.2 SIGNAL PROCESSING ALGORITHM AND SIMULATION EXAMPLES

Following from the above discussions, we will measure τ_1 at each recording instant n. First, the stationary component is removed by averaging all the delay estimates and then subtracting this average from the individual delay value. The observed data can then be represented as,

$$x[n] = B_1 \sin(2\pi \hat{f}_0 n + \psi) + B_2 \cos(4\pi \hat{f}_0 n + \varphi_t + \varphi_r) + w[n], \quad n = 1, 2, \ldots Q$$

$$(5.8)$$

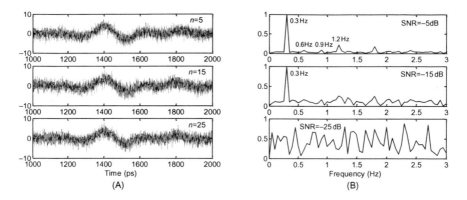

Figure 5.3 (A) Received signals at three recording instants and (B) periodograms for three different SNR values.

where $\hat{f}_0 = f_0 T_s$ is the normalized frequency and $B_1 > B_2$ as discussed in the previous section. The amplitudes B_1, B_2, and phases ψ, φ_t, and φ_r are unknown deterministic parameters. $w[n]$ is assumed to be white Gaussian noise, which accounts for delay estimation errors in the analysis. Subsequently, the frequency estimate is obtained by maximizing the periodogram,

$$I(\hat{f}) = \frac{1}{Q} \left| \sum_{n=1}^{Q} x[n] \exp(-j2\pi \hat{f} n) \right|^2 \tag{5.9}$$

over $\hat{f}(0 < \hat{f} < 0.5)$. Note that this actually corresponds to the maximum likelihood estimator of a single sinusoid embedded in white Gaussian noise.

In the numerical examples, it is assumed that the model parameters are the same as those in the previous section. Furthermore, the antenna heights are sinusoidal with $f_0 = 0.3$ Hz. The pulse repetition time $T_s = 100$ ms and a total number of 200 samples are collected to estimate the respiration rate (i.e., $Q = 200$). Figure 5.3A shows the signals received at the Rx at three different recording instants: $n = 5$, $n = 15$, and $n = 25$, where the signal-to-noise ratio (SNR) is equal to -5 dB. The subtractive deconvolution algorithm [49] is then applied to extract the delay information of various pulse echoes. With the knowledge of arrival time of the pulse reflected from the air–skin interface, Eq. (5.9) can be applied to identify the breathing speed. Figure 5.3B illustrates the periodograms for three different SNRs. When SNR = -5 dB, both the fundamental component at 0.3 Hz and higher harmonics are clearly visible. As SNR reduces to -15 dB, only the fundamental component can be identified. If we further decrease the SNR to -25 dB, the system performance deteriorates severely and an accurate estimate cannot be achieved. In such a case, a longer data length Q is required to recover the breathing rate.

CHAPTER 6

Conclusions

Wireless BANs are powered by the emergence of small and lightweight wireless systems such as Bluetooth enabled devices and PDAs. Antennas are essential part of any WBAN systems and due to varying requirements and constraints, careful considerations of their design and deployment are needed. This chapter introduced wireless BANs and their applications in consumer applications, military, and healthcare systems. The main requirements and features of wearable antennas were presented with regards to design and implementation issues. As inseparable part of the whole communication system, specifically in WBAN, the influence of different antenna parameters and types on the radio propagation channel is of great significance, specifically when designing antennas for wearable personal technologies. A case study of compact wearable antenna used in sensors aimed at healthcare applications was presented. Antenna's performance was investigated numerically with regard to impedance matching, radiation patterns, gain and efficiency. The compact small size of the sensor made it susceptible to variable changes caused by the human body and movements, specifically radiated power, efficiency, and front to back ratio of radiated energy.

Thereafter, the potential benefits and limitations of cooperative networks as a means of augmenting the reliability in body-wearable sensor were discussed. These trade-offs have been quantified for a sensor network prototyping a real-world platform for continuous ECG healthcare monitoring. For a packet delivery ratio >0.9, the multi-hop (MH) scheme can provide the network with a margin gain up to 14 dB, while resulting in an energy demand up to 30.7% higher and an average sensors r 20% lower than a single-hop (SH) scheme. The network lifetime of the MH scheme ranges from 27% to 45% of the SH lifetime in cases of 0 and 20 dB margins, respectively. This work was a first exercise step in assessing reliability and lifetime trade-off with real-world platforms for body area sensor networks. Follow-up studies will address wireless ECG emulators with higher number of sensors employing ultra-low power chipsets in different specific health

Co-operative and Energy Efficient Body Area and Wireless Sensor Networks for Healthcare Applications.
DOI: http://dx.doi.org/10.1016/B978-0-12-800736-5.00012-X

monitoring environments, such as critical care in hospitals, aged care, or athlete monitoring.

Finally, the use of UWB-radar-based wearable sensors in noncontact monitoring of respiration rates was demonstrated. Based on the proposed propagation model, it was shown that the delay information of the first echo conveys useful information on the breathing signal. Critical system design parameters have been identified through a sinusoidal approximation of the antenna heights. With a simple signal processing algorithm, it was verified via simulation that the respiration rate could be accurately measured for very low SNR values. Specifically, the influence of human body/cloth movements on the estimation accuracy was investigated. The sensor system could be used for other medical applications, such as aortic pressure measurement and cardiac monitoring, and be integrated into future cooperative, low-energy-consumption body area, and wireless sensor networks.

REFERENCES

[1] MIT Media Lab. <http://www.media.mit.edu/wearables/Communications/News/Pages/Global-Wireless-Subscriptions-Reach-5-Billion.aspx>. Accessed on 10 July 2011.

[2] Hall PS, Hao Y, Ito K. Guest editorial for the special issue on antennas and propagation on body-centric wireless communications. IEEE Trans Antenna Propag 2009;57(4):834–6.

[3] Hall PS, Hao Y. Antennas and propagation for body-centric wireless communications. MA, USA: Artech House; 2006.

[4] Allen B, Dohler M, Okon E, Malik WQ, Brown AK, Edwards D. UWB antenna and propagation for communications, radar and imaging. England, UK: John Wiley and Sons; 2007.

[5] Alomainy A. Antennas and radio propagation for body-centric wireless communications, PhD thesis, Queen Mary, University of London; 2007.

[6] Abbasi QH. Radio channel characterisation and system-level modelling for ultra wideband body-centric wireless communications, PhD thesis, Queen Mary, University of London; 2011.

[7] Sani A. Modeling and characterisation of antenna and propagation for body-centric wireless communications, PhD thesis, Queen Mary, University of London; 2010.

[8] An Internet Resource for the calculation of the dielectric properties of body tissues, Institute for Applied Physics, Italian National Research Council, <http://niremf.ifac.cnr.it/tissprop/>, last accessed on 10 January 2012.

[9] Gabriel C, Gabriel S. Compilation of the dielectric properties of body tissues at RF and microwave frequencies, <http://www.brooks.af.mil/AFRL/HED/hedr/reports/dielectric/Title/Title.html>; 1999.

[10] Alomainy A, Hao Y, Davenport DM. Parametric study of wearable antennas with varying distances from the body and different on-body positions, IET seminar on antennas and propagation for body-centric wireless communications, The Institute of Physics, London, UK; 24 April 2007.

[11] Alomainy A, Hao Y, Pasveer WF. Numerical and experimental evaluation of a compact sensor antenna performance for healthcare devices. IEEE Trans Biomed Circuits Syst 2007;1(4).

[12] Balanis C. Antenna theory analysis and design. Canada: John Wiley & Sons; 1997.

[13] Stutzman WL, Thiele GA. Antenna theory and design. John Wiley & Sons; 1998.

[14] Suma MN, Bybi PC, Mohanan P. A wideband printed monopole antenna for 2.4-GHz WLAN applications. Microw Opt Technol Lett 2006;48(5):871–3.

[15] Chen H-D, Chen J-S, Cheng Y-T. Modified inverted-L monopole antenna for 2.4/5 GHz dual-band operations. Electron Lett 2003;39(22):1567–8.

[16] Shum KM, Chan CH, Luk KM. Design of small antennas for joypad applications. In: IEEE international workshop on antenna technology small antennas and novel metamaterials. New York, USA; March 6–8, 2006. p. 168–71.

[17] Wireless USB Antenna Design Layout Guidelines-AN5032, Application Notes, Cypress Semiconductor Corporation, <http://www.cypress.com>; 2005.

[18] Jan J-Y, Tseng L-C, Chen W-S, Cheng Y-T. Printed monopole antennas stacked with a shorted parasitic wire for Bluetooth and WLAN applications. In: IEEE 2004 antennas and propagation society international symposium, vol. 3. Monterey, California, USA; June 2004. p. 2607–10.

[19] Chipcon CC2420 transceiver chip, 2.4 GHz IEEE 802.15.4/ZigBee-ready RF transceiver, <http://www.chipcon.com/files/CC2420_Data_Sheet_1_4.pdf>.

[20] Internet resource on IEEE std. 802.15.4—2003: Wireless Medium Access Control (MAC) and Physical Layer (PHY); <http://standards.ieee.org/about/get/802/802.15.html>. Accessed on 10 September 2011.

[21] Otto C, Milenkovic A, Sanders C, Jovanov E. System architecture of a wireless body area sensor network for ubiquitous health monitoring. J Mobile Multimedia 2006;1(4):307—26.

[22] Jovanov E, Milenkovic A, Otto C, De Groen PC. A wireless body area network of intelligent motion sensors for computer assisted physical rehabilitation. J Neuroeng Rehabil 2005;2(6).

[23] IEEE Standard document on Channel Model for Body Area Network (BAN) provided by IEEE P802.15 working group for Wireless Personal Area Networks (WPANs). Final document of the IEEE802.15.6 Channel Modeling Subcommittee. It provides how channel model should be developed for body area network. <https://mentor.ieee.org/802.15/file/08/15-08-0780-09-0006-tg6-channel-model.pdf>, last accessed Dec 2011.

[24] Strömmer E, Hillukkala M, Ylisaukkooja A. Ultra-low power sensors with near field communication for mobile applications presented at 2007 international conference on wireless network. Las Vegas, NV; 2007.

[25] Sagan D. RF integrated circuits for medical applications: meeting the challenge of ultra low power communication. San Diego, CA: Ultra-Low-Power Communications Division, Zarlink Semiconductor; 2005.

[26] Falck T, Baldus H, Espina J, Klabunde K. Plug 'n play simplicity for wireless medical body sensors. Mobile Netw Appl 2007;12(2—3):143—53.

[27] Mikami S, Matsuno T, Miyama M, Yoshimoto M, Ono H. A wireless-interface soc powered by energy harvesting for short range data communication. 2005 IEEE Asian solid-state circuits conference proceedings. Hsinchu, Taiwan; 2005. p. 241—44.

[28] Moerman I, Blondia C, Reusens E, Joseph W, Martens L, Demeester P. The need for cooperation and relaying in short-range high path loss sensor networks. In: Proceedings of the 2007 IEEE international conference on sensor technologies and applications. Washington, DC; 2007 p. 566—71.

[29] Chen Y, Teo J, Lai JCY, Gunawan E, Low KS, Soh CB, et al. Cooperative communications in ultra-wideband wireless body area networks: channel modelling and system diversity analysis. IEEE J Sel Areas Commun 2009;27(1):5—16.

[30] Latre B, Braem B, Moerman I, Blondia C, Reusens E, Joseph W, et al. A low-delay protocol for multihop wireless body area networks. In: Proceedings of the fourth international conference. MobiQuitous; 2007. p. 1—8.

[31] Braem B, Latré B, Benoît M, Blondia.C, Demeester P. The wireless autonomous spanning tree protocol for multi-hop wireless body area networks. Presented at 2006 third annual international conference on mobile and ubiquitous systems. San Jose, CA; 2006.

[32] Su-Ho S, Gopalan SA, Seung-Man C, Ki-Jung S, Jae-Wook N, Jong-Tae Park P. An energy-efficient configuration management for multi-hop wireless body area networks. In: Third IEEE international conference broadband network multimedia technology. Beijing, China; 2010. p. 1235—239.

[33] Djenouri E, Balasingham I. New QoS and geographical routing in wireless biomedical sensor networks. In: Proceedings of the sixth international conference on broadband communications, networks, and systems. Madrid (Spain); 2009, p. 1—8.

[34] Felemban E, Lee C-G, Ekici. E. MMSPEED: multipath multi-speed protocol for QoS guarantee of reliability and timeliness in wireless sensor networks. IEEE Trans Mobile Comput 2006;5(6):738—54.

[35] Razzaque A, Alam MM, Or-Rashid M, Hong CS. Multi-constrained QoS geographic routing for heterogeneous traffic in sensor networks. IEICE Trans Commun 2008;91B (8):2589–601.

[36] Chipara O, He Z, Xing G, Chen Q, Wang X, Lu C, et al. Real-time power aware routing in sensor networks. In: Proceedings of the IEEE 14th international workshop on quality of service. New Haven, Connecticut, USA; 2006. p. 83–92.

[37] Heinzelman WR, Chandrakasan A, Balakrishnan H. Energy-efficient communication protocol for wireless microsensor networks. In: Proceedings of the 33rd annual Hawaii international conference on system sciences, vol.2. Island of Maui, Hawaii, USA; 2000. p. 10, 4–7.

[38] ZigBee RF transceiver datasheet. Available from: <http://www.ti.com/lit/ds/symlink/cc2420.pdf>.

[39] Sentilla webpage. Available from: <http://www.sentilla.com/blogs/2008/05/sentilla-announces-worlds-smal.php>.

[40] Fan Y, Songwu L, Lixia Z. A scalable solution to minimum cost forwarding in large sensor networks. In: Proceedings of the 10th international conference computer communications and networks. Scottsdale, AZ, USA; 2001. p. 304–09.

[41] Holland MM, Aures RG, Heinzelman WB. Experimental investigation of radio performance in wireless sensor networks. Presented at 2nd IEEE workshop on wireless mesh networks. Reston, Virginia USA; 2006, p.140–50.

[42] Madan R, Lall S. Distributed algorithms for maximum lifetime routing in wireless sensor networks. IEEE Trans Wireless Commun 2006;5(8):2185–93.

[43] Bradie B. Wavelet packet-based compression of single lead ECG. IEEE Trans Biomed Eng 1996;43(5):493–501.

[44] Sha K, Shi W. Modeling the lifetime of wireless sensor networks. Sensor Lett 2005;3:1–10.

[45] Michahelles F, Wicki R, Shiele B. Less contact: heart-rate detection without even touching the user. In: Proceedings of the international symposium wearable computers, vol. 1. Arlington VA, USA; 2004. p. 4–7.

[46] Staderini EM. UWB radars in medicine. IEEE AESS Syst Mag 2002;17:13–8.

[47] Klemm M, Troester G. Textile UWB antennas for wireless body area networks. IEEE Trans Antennas Propag 2006;54:3192–7.

[48] Gabriel S, Lau RW, Gabriel C. The dielectric properties of biological tissues: III. Parametric models for the dielectric spectrum of tissues. Phys Med Biol 1996;41:2271–93.

[49] Chen Y, Gunawan E, Low KS, Wang S-C, Soh CB, Thi LL. Time of arrival data fusion method for two-dimensional ultrawideband breast cancer. IEEE Trans Antennas Propag 2007;55(10):2852–65.

CPSIA information can be obtained at www.ICGtesting.com
Printed in the USA
BVOW04s2019180514

353791BV00011B/93/P